PLANTS
AND
US

Angela Royston

Heinemann
LIBRARY

First published in Great Britain by Heinemann Library
Halley Court, Jordan Hill, Oxford OX2 8EJ
a division of Reed Educational and Professional Publishing Ltd.

Heinemann is a registered trademark of Reed Educational & Professional Publishing Limited.

OXFORD   MELBOURNE   AUCKLAND
JOHANNESBURG   BLANTYRE   GABORONE
IBADAN   PORTSMOUTH NH   CHICAGO

© Reed Educational and Professional Publishing Ltd 1999
The moral right of the proprietor has been asserted.

Designed by AMR Ltd.
Printed and bound in Hong Kong/China by South China Printing Co. Ltd.

03 02 01 00 99
10 9 8 7 6 5 4 3 2 1

ISBN 0 431 00201 0

British Library Cataloguing in Publication Data

Royston, Angela
  Plants and us.– (Plants)
  1. Plants, Useful – Juvenile literature
  I. Title
  581.6'3

  ISBN 0 431 00201 0

Acknowledgements
The Publishers would like to thank the following for permission to reproduce photographs:
Ardea: A Paterson p21; Liz Eddison: ppl3, 16; Garden and Wildlife Matters: pp7, 12, 26, 27, J Hoare p20, S North p8; Chris Honeywell: pp11, 14, 18, 22, 28, 29; Oxford Scientific Films: J McCammon p24, Oxapia p9; Science Photo Library: J Howard p23; The Stock Market: p6; Tony Stone Images: W Curtis p15, N Dolding p17, M Gowan p10, G Haling p19, R Torrez p5, P Tweedle p4; Trip: J Hurst p25.

Cover photograph: Paul Chesley, Tony Stone Worldwide

The Publishers would like to thank Dr John Feltwell of Garden Matters for his comments in the preparation of this book.

Every effort has been made to contact copyright holders of any material reproduced in this book. Any omissions will be rectified in subsequent printings if notice is given to the Publisher.

Any words appearing in bold, **like this**, are explained in the Glossary.

Contents

We all need plants

All animals and people rely on plants for food. Even animals that don't eat plants eat animals that do. This giraffe is eating the leaves of a thorny bush.

People grow fields of corn and other food plants. They also use plants for other things. This house is built of wood from trees.

Air fit to breathe

The air contains an important gas called **oxygen**. All living things breathe in oxygen and breathe out another gas, called **carbon dioxide**.

During the day, plants also take in
carbon dioxide through their leaves
and turn it into oxygen. So plants
keep the air stocked with vital oxygen.

Vegetables

We eat plants because they contain
vitamins and **minerals** that our
bodies need to stay healthy. Vegetables
are **roots**, **stems**, leaves or **flowers**.

Carrots and potatoes are swollen roots. We eat the leaves of cabbages and lettuces, and the flowers of cauliflower and broccoli.

A field of grain

Farmers all around the world grow fields of rice or wheat. This farmer is **harvesting** rice by hand. We eat the **seeds**, called grains.

When wheat is cut, the grains are collected and **ground** into a powder called flour. Flour is used to make bread, pasta and cakes.

Fruit and nuts

The fruit is the part of the plant which contains **seeds** that could grow into new plants. Apples, oranges, peaches and strawberries are all fruits.

Each fruit has one or more seeds
surrounded by sweet, juicy flesh. Nuts
are seeds too, but they are surrounded
by a hard shell.

Drinks

Apart from milk and water, most of
what we drink comes from plants.
Some fruits are squeezed to make juice.

The leaves of tea plants are picked by hand before they are dried and made into tea. Coffee beans are roasted and **ground** before they are used.

Spices and herbs

We add herbs and spices to food to
make it taste better. The leaves of these
herbs have a strong taste and smell.

Spices are made from the **roots** and
bark of plants that grow mainly in
hot countries. Many spices are **ground**
into powder before they are sold.

Creams and perfumes

All of these **cosmetics** have been made from plants. The labels tell you which plants have been used.

Flowers with a strong, sweet smell are made into perfumes. These lavender flowers may be used to perfume soap or talcum powder.

Medicine

In the past, most medicines came from plants. Today plants are still used to treat some illnesses. This little rosy periwinkle helps to treat leukaemia, a disease which affects people's blood.

The **bark** of the cinchona tree contains quinine, a drug which is used to treat an illness called malaria, which is common in some hot countries.

Wood and paper

All of these things are made from wood. The table is wooden too. Some kinds of trees are specially grown so that we can use their wood.

The wood of some trees is mashed down and made into huge rolls of paper for newspapers, books, packaging and other things.

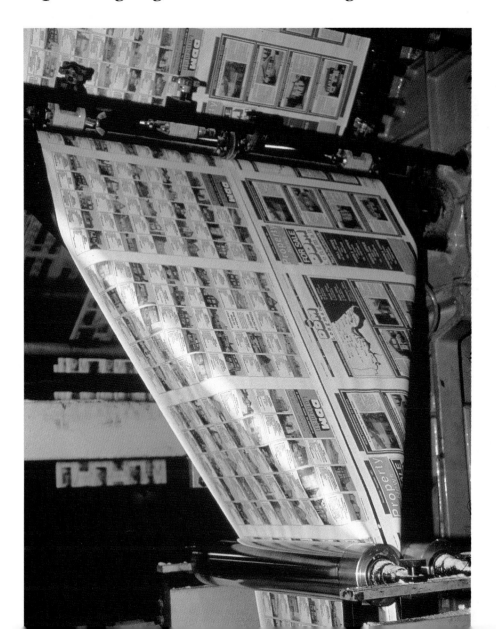

Clothes

Plants can be made into clothes. These fluffy **seeds** of cotton are spun into thread and woven into cotton cloth.

Many plants have very strong colours and these are used to make dyes. These clothes are being dyed several different colours.

Gardens and parks

Plants improve our lives in other ways too. Gardens and parks give us somewhere peaceful to relax and enjoy ourselves.

These **rainforest** plants are growing in a **national park** in Central America. We must protect all plants, not just the ones we use today.

Changing colour

You can use red cabbage to make a
dye, but you must ask an adult to help
you. Chop up the red cabbage and
put it in a pan of water with a small
piece of white cotton cloth.

Ask the adult to boil the pan on top
of the cooker for about half an hour,
making sure that the water does not
boil dry. Let the water cool. What
colour is the cloth now?

Plant map

a strawberry plant

leaf

flower

fruit

stem

roots

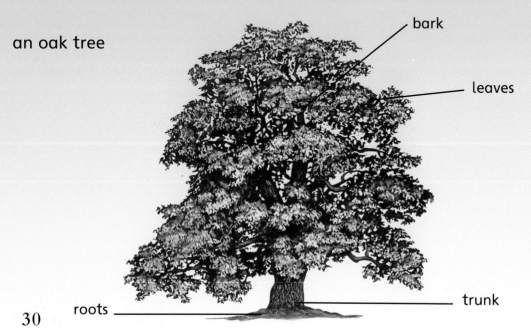

an oak tree

bark

leaves

trunk

roots

Glossary

bark	tough outer layer that protects the trunk of a tree
carbon dioxide	a gas which is made when living things breathe out and when fuel is burnt
cosmetics	creams, lotions and powders which people use on their skin and hair
flower	the part of a plant which makes new seeds
ground	broken up into a powder
harvesting	gathering crops, like wheat, when they are fully grown
minerals	substances found in the earth which plants and animals need to stay healthy
national park	area of land which has been put aside for plants or animals to live without being disturbed by people
oxygen	a gas which all living things need to breathe to survive
rainforest	rainy place where many trees and plants grow together
roots	parts of a plant which take in water, especially from the soil
seed	a seed contains a tiny plant before it begins to grow and a store of food
stem	the part of a plant from which the leaves and flowers grow
vitamins	special kind of foods which animals and people need to stay healthy

Index